Diabetes Control

The Complete Guide on Type 1 & Type 2 Diabetes, Pre-Diabetes, Gestational Diabetes, Signs and Symptoms, Causes and Treatments.

Doctor Jane A. McCall

Table of Contents

Diabetes Control 1

INTRODUCTION 5

CHAPTER 1 6

    WHAT IS DIABETES? 6

CHAPTER 2 9

    WHAT IS PRE-DIABETES? 9

        Symptoms of Pre-Diabetes 10

        Causes and Risk Factors of Pre-Diabetes 11

        Diagnosis of Pre-Diabetes 16

        How to Treat Pre-Diabetes 20

CHAPTER 3 25

    WHAT IS TYPE 1 DIABETES 25

        Causes of Type 1 Diabetes 27

        Common Symptoms and Signs of Type 1 Diabetes. 29

        Life with Type 1 Diabetes 31

CHAPTER 4 33

    WHAT IS TYPE 2 DIABETES 33

        Causes of Type 2 Diabetes 35

        Common Diabetes Symptoms Associated With Type 2 Diabetes. 37

CHAPTER 5 40
WHAT IS GESTATIONAL DIABETES 40
Maturity Onset Diabetes of The Young 42

CHAPTER 6 44
COMPLICATIONS OF BADLY CONTROLLED DIABETES 44
Symptoms Caused By Diabetes Complications 47
Symptoms Related To Nerve Damage (Neuropathy) 49
Skin-Related Diabetes Symptoms: 52
Eye-Related Diabetes Symptoms 54

CHAPTER 7 57
CONTROLLING DIABETES-TREATMENT IS EFFECTIVE AND IMPORTANT. 57
Natural Ways to Help Control Diabetes Symptoms 59

Recommendations 67
About the Author 71
Acknowledgments 72

Copyright © 2019 by Doctor Jane A. McCall

All rights reserved. No part of this publication may be reproduced, distributed, or transmitted in any form or by any means, including photocopying, recording, or other electronic or mechanical methods, without the prior written permission of the publisher, except in the case of brief quotations embodied in critical reviews and certain other non-commercial uses permitted by copyright law.

INTRODUCTION

Most people in the world are suffering from diabetes. Research has shown that millions of people in the United State, Asian Countries, Africa etc. Majority of people are undiagnosed and unaware of their condition.

There are lots of things that contribute to this (i.e causes) diabetes. This book is been written to reach out to each and every one all across the globe affected with diabetes.

In this book you will be enlightened on the causes related to diabetes, symptoms related as for those undiagnosed to know what they are up against and the best possible treatment concerning diabetes.

CHAPTER 1

WHAT IS DIABETES?

Diabetes, frequently known as diabetes mellitus, is a group of metabolic diseases in which the victim has excessive blood glucose (blood sugar), either due to the fact insulin production is inadequate, or due to the fact that the cells of the body's fail to respond well to insulin, or both. Patients with excessive (high) blood sugar will frequently experience polyuria (common urination), they may end up more and more thirsty (polydipsia) and increased hunger (polyphagia). If left untreated, diabetes can cause many complications. Acute complications can include diabetic ketoacidosis, hyperosmolar hyperglycemic state, or demise. Serious long-term complications encompass cardiovascular sickness, stroke, chronic kidney disorder, foot ulcers, and harm to the eyes. Diabetes is due to either the pancreas not producing enough insulin or the cells of the body not responding properly to the insulin produced. The fundamental types of diabetes mellitus:

- Type 1 DM (diabetes mellitus): results from the pancreas's failure to produce enough insulin. This form was formally referred to as "insulin-dependent diabetes mellitus" (IDDM) or "juvenile diabetes".

- Type 2 DM (diabetes mellitus): starts off with insulin resistance, a situation in which cells fail to reply to insulin properly. as the disorder progresses a loss of insulin may additionally expand. This shape was formerly called "non-insulin-dependent diabetes mellitus" (NIDDM) or "person-onset diabetes".

- Gestational diabetes: is the third fundamental form and this do occurs when pregnant women without a preceding history of diabetes have high blood sugar levels due to one factor or the other.

CHAPTER 2

WHAT IS PRE-DIABETES?

Pre-diabetes is the precursor stage before diabetes mellitus wherein not all the signs and symptoms required to diagnose diabetes are present, however blood sugar is abnormally high. This stage is frequently referred to as the "grey area." it is not always a disease likewise pre-diabetes is a "pre-analysis" of diabetes—you could consider it as a warning sign. It's while your blood glucose level (blood sugar level) is higher than what it ought to be, but it's not high enough to be taken into consideration as diabetes.

Pre-diabetes is a sign that you may develop Type 2 diabetes in case you don't make a few lifestyle modifications.

It's quite possible to prevent pre-diabetes from growing into type 2 diabetes. Eating healthy food, losing weight and staying at a healthy weight, and being bodily active can help you bring your blood glucose level reduced back into the normal stage.

Symptoms of Pre-Diabetes

Diabetes develops very progressively, so while you're in the pre-diabetes stages-when your blood glucose level is higher than it ought to be-you could no longer have any symptoms in any respect. You can, however, notice that:

- You're hungrier than normal

- You're dropping weight, in spite of ingesting more

- You're thirstier than normal

- You have to visit the rest room more frequently

- You're more worn-out than regular

All of these are normal signs and symptoms related to diabetes, so if you're in the early degrees of diabetes, you may note them.

Causes and Risk Factors of Pre-Diabetes

Pre-diabetes develops while your body starts to have hassle using the hormone insulin. Insulin is important to transport glucose-what your body uses for energy-into the cells through the bloodstream. In pre-diabetes, it shows that your body doesn't make enough insulin or it doesn't use it properly (that's referred to as insulin resistance).

If you don't have sufficient insulin or if you're insulin resistant, you can build up too much glucose for your blood, leading to a higher than ordinary blood glucose level and possibly pre-diabetes.

Researchers aren't sure what precisely causes the insulin process to go awry in some people. There are several risk elements, though, that make it much more likely that you'll develop pre-diabetes. Those are the identical danger factors related to the development of type 2 diabetes:

1. **Weight:** in case you're overweight (have a body mass index-a BMI-higher than 25), you're at an excessive risk for developing pre-diabetes. Most especially if you carry quite a few extra weight in your abdomen, you may develop pre-diabetes. The more fat cells can cause your body to end up more insulin resistant.

2. **Lack of physical activity:** this often goes hand-in-hand with being overweight. If you aren't physically active, you're much more likely to develop pre-diabetes.

3. **Family history:** pre-diabetes has a hereditary component. If someone in your own family has (or had) it, you're more likely to develop it.

4. **Race/ethnicity:** certain ethnic groups are much more likely to increase pre-diabetes, such as Africans-Americans, Hispanic Americans, Native Americans, and Asian Americans.

5. **Age:** the older you are, the more at risk you're for developing pre-diabetes. At age forty-five, your chance starts to rise, and after age sixty-five, your chance increases exponentially.

6. **Gestational diabetes:** if you develop diabetes as at the time you were pregnant, that will increase your risk for developing pre-diabetes afterward.

7. **Other health problems:** excessive blood pressure (hypertension) and high cholesterol (the "bad" LDL cholesterol) increase your hazard of having type 2 diabetes.

Polycystic ovary syndrome (PCOS) additionally raises the chance for pre-diabetes as it is associated with insulin resistance. In PCOS, many cysts form in your ovaries, and one possible reason is insulin resistance. When you have PCOS, which means you'll be insulin resistant and consequently at risk for developing pre- diabetes.

Diagnosis of Pre-Diabetes

Your doctor may need to check your blood glucose level in case you're overweight (have a body mass index-BMI-

of over 25) and when you have one or more of the risk factors indexed above.

Even if you are not overweight and don't have any of the risk factors, your health practitioner might also need to begin testing out your blood glucose level every three years beginning while you're forty-five. That's a smart factor to do because the risk of developing pre-diabetes (and consequently type 2 diabetes) increases with age. Due to the fact that there are so many possible complication of diabetes (e.g., heart problems and nerve issues), it's an awesome idea to be vigilant about detecting blood glucose abnormalities early.

To diagnose you with pre-diabetes, the doctor can run one of two tests-or he/she may decide to do both. The checks are:

Fasting plasma glucose test (FPG): you won't eat anything for 8 hours leading up to a FPG take a look at. That's why a FGP test is often done in the morning. The doctors exams your blood glucose level (blood sugar level) after drawing a small blood sample.

In case your blood glucose level is among 100 and 125mg/dl, you have pre-diabetes. You may hear the doctor use the word "impaired fasting glucose" or IFG, that is another term for pre-diabetes whilst it is identified with the fasting plasma glucose test.

In case your blood glucose level is above 126mg/dl with the FGP test, you may have diabetes.

Oral glucose tolerance test (OGTT): that is some other test used to diagnose pre-diabetes. The medical doctor will give you instructions on the way to prepare for the test, but you won't be able to eat anything for 8 hours earlier than the test; you'll be fasting. In that way, the oral glucose tolerance test, abbreviated OGTT, is much like the fasting plasma glucose test.

On the day of the test, the health practitioner will test your blood glucose level at the beginning of the appointment; that's known as your fasting blood glucose level. Then, you'll drink 75g of a completely sugary mixture. Two hours later, your blood glucose degree will be measured.

In case your blood glucose degree is between 140 and 199mg/dl two hours after ingesting the sugary combination, you have got pre-diabetes. You may hear the doctor use the word "impaired glucose tolerance" or IGT, which is another term for pre-diabetes while it's diagnosed with the OGTT.

If your blood glucose level is above 200mg/dl with the oral glucose tolerance test, you may have diabetes.

How to Treat Pre-Diabetes

The American diabetes association says that several lifestyle changes are powerful in preventing Type 2 diabetes once you've been diagnosed with pre-diabetes. Your doctor will walk you through what you need to change, however typical recommendations are:

1. **Eat properly:** a registered dietitian (RD) or licensed diabetes educator (CDE) will help you create a meal plan that's full of goods-for-you and good-for-your-blood-glucose-level meals. The intention of the meal plan is to control your blood glucose level and keep it in the healthy, normal range. Your meal plan will be made just for you, taking into account your overall health, physical activity, and what you like to eat.

2. **Exercise:** when you exercise, your body uses more glucose, so exercising can reduce your blood glucose level. In addition while you exercise, your body doesn't need much insulin to transport the glucose; your body becomes much less insulin resistant. Since your body isn't using insulin properly if you have pre-diabetes, a decrease insulin resistance is a very good thing.

And of cause, there are all of the conventional benefits of exercise: it can help you lose weight, keep your coronary heart healthy, make you sleep better, and even enhance your mood.

The American diabetes association recommends at the least a 150minutes of moderate activity per week—that's 30minutes five days a week. You could get that through sports consisting of walking, bike using, or swimming.

3. **Lose weight:** If you're overweight, you should get started on a weight loss as soon as you're diagnosed with pre-diabetes. Losing simply 5-10% of your weight can significantly reduce your hazard of developing Type 2 diabetes. The aggregate of consuming nicely and exercise more is a splendid way to lose weight-after which you maintain your new, healthful weight.

4. **Metformin:** for peoples that are at a very high risk of developing Type 2 diabetes after being diagnosed with pre-diabetes, the doctor may suggest a medication.

The American diabetes association says that metformin should be the handiest medication used to prevent type 2. It really works by retaining the liver from making greater glucose when you don't want it, thereby maintaining your blood glucose level in a better range.

Your health practitioner will keep a near watch on your blood glucose level, monitoring them to make sure that your pre-diabetes doesn't end up Type 2 diabetes. If needed, he or she may suggest adjustment changes (e.g., different diet or more exercise) to better control your blood glucose levels.

CHAPTER 3

WHAT IS TYPE 1 DIABETES

Type 1 diabetes is an autoimmune ailment wherein the immune gadget destroys cells inside the pancreas. The body does no longer produce insulin which will help to fight against any form of infection or disease. Normally, the disease first appears in adolescence or early adulthood. Type 1 diabetes used to be called juvenile-onset diabetes or insulin-based diabetes mellitus (IDDM), however the disease could have an onset at any age. Type 1 diabetes makes up round 5% of all instances of diabetes.

In Type 1 diabetes, the pancreas is unable to produce any

insulin, the hormone that controls blood sugar levels. Insulin production turns into inadequate for the control of blood glucose levels due to the sluggish destruction of beta cells inside the pancreas. This destruction progresses without been aware over time until the mass of those cells decreases. To the extent that the quantity of insulin produced is insufficient.

Type 1 diabetes generally appears in childhood or youth, however its onset is likewise possible in maturity.

While it develops later in lifestyles, type 1 diabetes can be mistaken for type 2 diabetes. If correctly diagnose, it's miles referred to as latent autoimmune diabetes of maturity.

Causes of Type 1 Diabetes

The gradual destruction of beta cells in the pancreas that sooner or later consequences within the onset of type 1 diabetes is the result of autoimmune destruction. The immune machine turning against the body's own cells is probably brought on by means of an environmental component uncovered to human beings who've a genetic susceptibility.

Despite the fact that the mechanisms of Type 1 diabetes etiology are uncertain, they're thought to contain the interaction of more than one factor which are as follows:

- Susceptibility genes - some of which might be carried by over 90% of patients with type 1 diabetes. A few populations - scandinavians and sardinians, for example - are much more likely to have susceptibility genes.

- Autoantigens - proteins thought to be launched or exposed at some point of ordinary pancreas beta cell turnover or Damage along with that caused by infection. The autoantigens set off an immune response resulting in beta cell Destruction.

- Viruses - coxsackievirus, rubella virus, cytomegalovirus, epstein-barr virus and retroviruses are amongst the ones which have been connected to type 1 diabetes.

- Eating regimen(Diet)- toddler exposure to dairy products, high nitrates in consuming water and low vitamin D intake have additionally been connected to the development of type 1 diabetes.

Common Symptoms and Signs of Type 1 Diabetes.

- Regularly feeling thirsty and having a dry mouth
- Changes on your urge for food, commonly feeling very hungry, occasionally even if you've currently eaten (this could also occur with weak point and problem Concentrating)
- Fatigue, feeling usually tired no matter slumbering and mood swings
- Blurred, worsening vision

- Gradual recuperation of skin wounds, common infections, dryness, cuts and bruises
- Unexplained weight changes, specifically dropping weight despite consuming the identical quantity (this occurs due to the body using alternative fuels stored in muscle and fat whilst releasing glucose in the urine)
- Heavy respiratory
- Potentially a loss of recognition
- Nerve damage that causes tingling sensations or ache and numbness in the limbs, feet and arms (more common among people with Type 2 Diabetes)

Life with Type 1 Diabetes

Type 1 diabetes continually requires insulin treatment and an insulin pump or day by day injections could be a lifelong requirement to keep blood sugar levels under control. The situation was known as insulin structured diabetes.

After the prognosis of type 1 diabetes, health care providers should help patients discover ways to self-monitor via finger stick testing, the signs of hypoglycemia, hyperglycemia and different diabetic complication. Most patients will also be taught how to regulate their insulin doses.

As with other forms of diabetes, vitamins and physical hobby and workout are vital factors of the way of life control of the sickness.

CHAPTER 4

WHAT IS TYPE 2 DIABETES

Type 2 diabetes is the most popular form of diabetes, accounting for over 90% of all diabetes cases.

The number of adults identified with diabetes within the US has risen notably inside the past 30 years, almost quadrupling from 5.5 million cases in 2000 to more than 21.3 million in 2017.

Type 2 diabetes was once called adult-onset diabetes and Noninsulin-structured diabetes mellitus (NIDDM), however the ailment may have an onset at any age, increasingly more along with early life.

Type 2 diabetes mellitus most commonly develops in

adulthood and is more likely to occur in folks that are overweight and physically inactive.

Not like Type 1 diabetes which presently cannot be prevented, a number of the risk factors for type 2 diabetes may be modified. For lots of people, consequently, it is viable to save you the circumstance. Signs and symptoms that signal the need for diabetes testing:

- Frequent urination

- Weight reduction

- Loss of energy

- Excessive thirst.

- Slow healing of cuts

- Numbness or tingling in hands and feet

- Itchy skin

Causes of Type 2 Diabetes

Insulin resistance is normally the precursor to type 2 diabetes - a situation in which more insulin than normal is needed for glucose to enter cells. Insulin resistance inside the liver results in more glucose production at the same time as resistance in peripheral tissues means that glucose uptake is impaired. The impairment stimulates the pancreas to make extra insulin however sooner or later the pancreas is unable to make sufficient to save your blood sugar levels from growing too excessive.

Genetics performs a part in Type 2 diabetes - relatives of people with the disease are at a higher risk, and the Prevalence of the condition is much higher in particular amongst local Americans, Hispanic and Asian human beings.

Obesity and weight gain are vital elements that lead to insulin resistance and Type 2 diabetes, with genetics, diet, exercise and way of life all playing an element. Body fat has hormonal effects on the impact of insulin and glucose metabolism.

Once type 2 diabetes has been identified, health care provider can help patients with a program of education and monitoring, including the way to spot the signs of hypoglycemia, hyperglycemia and other diabetic intricates.

As with other kinds of diabetes, nutrients, and bodily pastime and exercise are critical elements of the life-style Management of the situation.

Common Diabetes Symptoms Associated With Type 2 Diabetes.

Many people develop type 2 diabetes symptoms in midlife or in older age and gradually expand signs in stages, especially if the condition is going untreated and worsens. Type 2 diabetes signs and symptoms can consist of:

- Chronically dry and itchy skin
- Patches of dark, velvety skin inside the folds and creases of the body (normally in the armpits and neck). This is known as acanthosis nigricans.
- Common infections (urinary, vaginal, yeast and of the groin)
- Weight benefit, even without a change within the diet
- Ache, swelling, numbness or tingling of the hands and toes

- Sexual disorder, consisting of loss of libido, reproductive issues, vaginal Dryness and erectile dysfunction.

CHAPTER 5

WHAT IS GESTATIONAL DIABETES

Gestational diabetes mellitus (GDM) resembles Type 2 DM in several respects, regarding a mixture of enormously inadequate insulin secretion and responsiveness. It takes place in about 2–10% of all pregnancies and might improve or disappear after delivery.

But, after pregnancy about 5–10% of ladies with gestational diabetes are found to have diabetes mellitus, most normally Type 2.

Gestational diabetes is fully treatable, but calls for careful medical supervision throughout the pregnancy. Management may additionally encompass dietary changes, blood glucose tracking, and in a few instances, insulin can be required.

Though it may be temporary, untreated gestational diabetes can harm the fitness of the fetus or mother. Dangers to the baby include macrosomia (high birth weight), congenital coronary heart and central nervous system abnormalities, and skeletal muscle malformations.

Expanded tiers of insulin in a fetus's blood may additionally inhibit fetal surfactant production and cause respiratory distress syndrome. A high blood bilirubin stage may additionally result from red blood cell destruction. In intense instances, perinatal loss of life may occur, most commonly because of negative placental perfusion because of vascular impairment. labor induction may be indicated with reduced placental function. A caesarean section may be done if there is marked fetal distress or an increased hazard of damage related to macrosomia, along with shoulder dystocia.

Maturity Onset Diabetes of The Young

Maturity onset diabetes of the young (MODY) is an autosomal dominant inherited form of diabetes, due to several one of numerous single-gene mutations inflicting defects in insulin production. It's far drastically less not unusual than the 3 main types (Type 1, Type 2, and Gestational). The name of this sickness refers to early hypotheses as to its nature. Being due to a defective gene, this ailment varies in age at presentation and in severity in keeping with the precise gene disorder. Human beings with MODY regularly can manage it without the use of insulin.

CHAPTER 6

COMPLICATIONS OF BADLY CONTROLLED DIABETES

Below is a list of possible complications that may be due to badly managed diabetes:

- Eye complications - glaucoma, cataracts, diabetic retinopathy, and a few others.

- Foot complications - neuropathy, ulcers, and every now and then gangrene which can also require that the foot be amputated

- Pores and skin complications - human beings with diabetes are extra vulnerable to pores and skin infections and pores and skin disorders

- Coronary heart problems - including ischemic heart disease, whilst the blood supply to the coronary heart muscle is diminished

- High blood pressure - not unusual in humans with diabetes, which could enhance the threat of kidney disease, eye troubles, heart assault and stroke

- Intellectual fitness - out of control diabetes raises the risk of laid low with depression, anxiety and a few different intellectual issues

- Hearing loss - diabetes sufferers have a higher danger of developing hearing problems

- Gum sickness - there may be a miles higher prevalence of gum disorder amongst diabetes patients

- Gastroparesis - the muscle groups of the belly prevent running well

- Ketoacidosis - a aggregate of ketosis and acidosis; accumulation of ketone bodies and acidity inside the blood.

- Neuropathy - diabetic neuropathy is a kind of nerve harm that can cause several different issues.

- HHNS (Hyperosmolar Hyperglycemic Nonketotic Syndrome) - blood glucose degrees shoot up too excessive, and there aren't any ketones present inside the blood or urine. It is miles an emergency circumstance.

- Nephropathy - out of control blood strain can result in kidney disorder

- Pad (peripheral arterial disease) - symptoms may additionally consist of pain in the leg, tingling and every now and then issues strolling properly

- Stroke - if blood stress, levels of cholesterol, and blood glucose stages are not managed, the danger of stroke will increase notably erectile disorder - male impotence.

- Infections - people with badly managed diabetes are much more vulnerable to infections.

- Recovery of wounds - cuts and lesions take a great deal longer to heal

Symptoms Caused By Diabetes Complications

It's possible to experience many complications from diabetes that cause others, usually extra drastic and harmful signs. That is why early detection and remedy of diabetes is so critical — it is able to substantially decrease the threat of growing complications like nerve harm, cardiovascular troubles, pores and skin infections, similarly weight advantage/infection and extra.

How in all likelihood are you to experience complication? Numerous elements impact whether or not you will broaden worsened signs or complications because of diabetes, including:

- How well you manage blood sugar stages
- Your blood pressure levels
- How long you've had diabetes
- Your circle of relatives history/genes your

- Lifestyles, such as your weight loss program, exercising routine, stress levels and sleep.

Symptoms Related To Nerve Damage (Neuropathy)

A complete half of all people with diabetes will develop a few form of nerve damage, mainly if it is going out of control for decades and blood glucose levels remain abnormal. There are several one of a kind sorts of nerve damage as a result of diabetes that could cause numerous symptoms: peripheral neuropathy (which affects the feet and hands), autonomic neuropathy (which influences organs like the bladder, intestinal tract and genitals), and several other forms that cause damage to the spine, joints, cranial nerves, eyes and blood vessels.

Signs and symptoms of nerve harm due to diabetes can include:

- Tingling within the feet, described as "pins and needles"
- Burning, stabbing or taking pictures pains in the feet and hands
- Sensitive pores and skin that feels very warm or cold
- Muscle aches, weakness and unsteadiness
- Rapid heartbeats
- Trouble sound asleep

- Changes in perspiration
- Erectile disorder, vaginal dryness and lack of orgasms as a result of nerve damage across the genitals
- Carpal tunnel syndrome
- Proneness to accidents or falling
- Changes in the senses, consisting of listening to, sight, taste and odor
- Problem with normal digestion, including frequent belly bloating, constipation, diarrhea, heartburn, nausea, vomiting.

Skin-Related Diabetes Symptoms:

One of the regions affected most and quickest by diabetes is the skin. Diabetes signs on the skin may be some of the most easy to recognize and earliest to show up. Some of the ways that diabetes influences the pores and skin is by causing terrible flow, sluggish wound restoration, reduced immune feature, and itching or dryness. This makes yeast infections, bacterial infections and different skin more easy to increase and harder to do away with. Skin troubles induced by diabetes consist of:

- Rashes/infections which can be occasionally itchy, hot, swollen, crimson and painful

- Bacterial infections (which includes vaginal yeast infections bacteria, also referred to as staph)

- Styles in the eyes and eyelids

- Acne

- Fungal infections (inclusive of candida symptoms that affect the digestive tract and fungus in pores and skin folds, together with across the nails, below the breasts, between the hands or toes, inside the mouth, and across the genitals) Jock itch, athlete's foot and ringworm

- Dermopathy

- Necrobiosis lipoidica diabeticorum

- Blisters and scales, particularly round infections

- Folliculitis (infections of hair follicles)

Eye-Related Diabetes Symptoms

Having diabetes is one in all the most important factors for developing eye troubles and even imaginative and prescient loss/blindness. Human beings with diabetes have a higher danger of blindness than human beings without diabetes, however most only developed minor problem that can be handled earlier than they get worse. Diabetes affects the outer, difficult membrane part of the eyes; the front component, which is obvious and curved; the cornea/retina, which consciousness light; and the macula.

Signs and symptoms of diabetes related to imaginative and prescient/eye health can encompass:

- Diabetic retinopathy (a term for all disorders of the retina as a result of diabetes, such as nonproliferative and proliferative retinopathy)
- Nerve damage to the eyes
- Cataracts
- Glaucoma
- Macular degeneration
- Seeing spots, vision loss and even blindness

One of the areas of the eyes maximum impacted by diabetes is the macula, that is specialized for seeing nice details and permitting us to see with sharp vision. Problems with blood drift making its way from the retina to the macula leads to glaucoma, which is 40%more likely to occur in humans with diabetes than in healthy humans. Danger for glaucoma goes up the longer a person has had diabetes and also the older a person turns into.

CHAPTER 7

CONTROLLING DIABETES-TREATMENT IS EFFECTIVE AND IMPORTANT.

All varieties of diabetes are treatable. Diabetes Type 1 lasts an entire life; there may be no known cure. Type 2 normally lasts a lifetime, however, some people have managed to put off their symptoms without medicinal drug, through a combination of exercise, diet and frame weight control.

Gastric bypass surgery can reverse type 2 diabetes in a high percentage of sufferers. Inside three to five years the disease recurs in about 21% of those that did the surgery.

"The recurrence rate was greatly stimulated by means of longstanding records of Type 2 diabetes before the surgical operation. This suggests that early surgical intervention in the overweight, diabetic population will improve the durability of remission of Type 2 diabetes."

Sufferers with Type 1 are treated with regular insulin injections, as well as a special weight-reduction plan and exercise. Sufferers with type 2 diabetes are usually dealt with capsules, workout and a unique diet, but now and again insulin injections are also required. If diabetes is not correctly controlled the patient has a significantly better chance of growing complications.

Natural Ways to Help Control Diabetes Symptoms

Diabetes is a serious condition that incorporates many risks and signs and symptoms; however the desirable information is it could be controlled with correct treatment and lifestyle modifications. A high percentage of humans with Type 2 diabetes are able to reverse and manipulate their diabetes signs completely and clearly through enhancing their diets, tiers of bodily interest, sleep and pressure stages. And even though Type 1 diabetes is tougher to deal with and manipulate, complications can also be decreased by taking the same steps. One of the pleasant things to do to prevent diabetes symptoms from worsening is to educate yourself approximately how diabetes forms and worsens.

1. Keep Up with Regular Checkups

Many people with complications of diabetes won't have noticeable symptoms (for example, nonproliferative retinopathy, which may cause vision loss or gestational diabetes at some stage in pregnancy). This makes it certainly crucial that you get looked at by using your doctor frequently to display your blood sugar levels, development, eyes, skin, blood strain tiers, weight and heart.

To be certain that you don't put yourself at a higher risk for coronary heart problem, work with your medical doctor to ensure you hold near normal blood pressure, and triglyceride (lipid) levels. Ideally, your blood pressure shouldn't move over 130/80. You must also attempt to preserve a wholesome weight and decrease inflammation in popular. The quality manner to do that is to devour an unprocessed, wholesome weight loss plan as well as exercising and sleep nicely.

2. Eat a Balanced Diet and Exercise

As part of a healthy diabetes weight loss plan, you could help maintain your blood sugar inside the normal variety via eating unprocessed, complete ingredients and averting things like delivered sugars, trans fats, processed grains and starches, and conventional dairy products.

Physical state of being inactive and weight problems (Obesity) are strongly associated with the development of type 2 diabetes, which is why exercise is important to control symptoms and lower the risk for headaches, inclusive of heart ailment. The countrywide institute of fitness states that humans can sharply lower their risk for diabetes through dropping weight via regular bodily pastime and a food plan low in sugar, refined fat and excess energy from processed foods.

3. Control Blood Sugar to Help Stop Nerve Damage

The quality manner to help prevent or delay nerve damage is to closely adjust your blood sugar levels. If you suffer from digestive issues due to nerve damage affects your digestive organs, you may benefit from taking digestive enzymes and supplements like magnesium which could help loosen up muscle groups, improve gut health and control signs. Other problems like hormonal imbalances sexual dysfunctions and trouble dozing also can be significantly decreased when you enhance your eating regimen, nutrient consumption, pressure levels and condition overall.

4. Help Protect and Treat the Skin

Humans with diabetes tend to have extra bacterial, fungal and yeast infections than healthy human beings do. When you have diabetes, you can help prevent pores and skin problems by means of dealing with your blood sugar ranges, practicing proper hygiene and treating skin obviously with things like *vital oils*.

Medical practitioners additionally endorse you limit how frequently you shower whilst your skin is dry, use natural and slight products to clean your skin (in place of many harsh, chemical products sold in most shops), moisturize daily with something mild like *coconut oil for skin*, and avoid burning your skin inside the sun.

5. Safeguard the Eyes

peoples that preserve their blood sugar levels closer to normal are less likely to have vision-related issues or at the least more likely to enjoy milder signs and symptoms. Early detection and appropriate follow-up care can save your Imaginative and prescient. To assist decrease the hazard for eye-related problems like moderate cataracts or glaucoma, you need to have your eyes checked at least one to two instances every year. Staying physically energetic and preserving a healthy weight loss program can save you or delay vision loss by way of controlling blood sugar, plus you need to additionally put on sun shades whilst in the solar. If your eyes become extra damaged over time, your physician may additionally propose you get hold of a lens transplant to hold

imaginative and prescient.

Recommendations

1) [How and Where to Buy Viagra Online Safely, Legally and Cheap](http://getbook.at/viagraonline): The Secret Behind How To Buy Viagra Online Safely Without A Prescription (With List Of Best Place To Buy Viagra Online) http://getbook.at/viagraonline

2) [Viagra & Sildenafil: Uses, Dosage, Side Effects and Risks Information](http://getBook.at/viagra): The Secret Guide Behind How To Buy Viagra Online Safely, Cheap and Legally (With Best Online Pharmacy for Generic Viagra) http://getBook.at/viagra

3) Erectile Dysfunction (ED): Symptoms & Causes, Diagnosis, Treatment Online, And More Using Viagra Without A Prescription (Including Where

To Buy Viagra, Cialis, Levitra etc Drugs Cheap & Safely Online
http://getBook.at/erectile

4) Innovative Visualisation: The Power of Mind Perception -- GET MORE DONE THROUGH MIND MANIPULATION, INCENTIVES, PSYCH TRICKS AND MORE
http://getBook.at/innovative

5) Natural Healing and Remedies Cyclopedia: Complete solution with herbal medicine, Essential oils natural remedies and natural cure to various illness. (The answer to prayer for healing)
http://getBook.at/naturalhealing

6) 100 BEST CAT WELLNESS FOOD, DIET & RECIPES: The hidden healing power diet for cat kidney problems, cat weight-loss, & pregnant cat

diet; including recipes for all cat diseases and illness http://myBook.to/catfood

7) The Brain, Mind and Memory Therapy: The Science of embracing Change, Boosting Brain Power, Increasing Your Energy and Mental Strength.
http://getBook.at/brainbook

8) What Wikipedia Can't Tell You About Achieving Your Goals: Why your objective setting never works out the way you plan
http://getBook.at/wikipedia

9) The First Year From Childbirth and beyond: Inside-out Information on what to expect the first year and beyond early childhood for mothers and fathers made simple
http://getBook.at/childbirth

We love Testimonies, and we want to know how thus our publications have been of immense help to you. And please consider writing to us at www.engolee.com

Follow us on Social media at:

Website: www.engolee.com

Facebook Page: www.facebook.com/engolee

Twitter Page: www.twitter.com/engolee

About the Author

Dr. Jane A. McCall is a willing Health Researcher who is committed to blessing human race. She has developed a series of fabulous and highly effective healthful strategies and exercise programs. She applies her encyclopaedic knowledge and astonishing perception to analyze the background and underlying causes of various diseases affecting people in the world and then designs individualized and totally effective strategies to attain the desired results in solving human related problem with diseases. Jane is totally committed to helping the world discover their ideal expression of complete wellbeing.

Acknowledgments

The Glory of this book success goes to God Almighty and my beautiful Family, Fans, Readers & well-wishers, Customers and Friends for their endless support and encouragements.

www.ingramcontent.com/pod-product-compliance
Lightning Source LLC
Chambersburg PA
CBHW021506210526
45463CB00002B/909